Gaia Warnings

Palewell Press

Gaia Warnings

Poems by Philip Burton

Gaia Warnings

First edition 2021 from Palewell Press,
www.palewellpress.co.uk

Printed and bound in the UK

ISBN 978-1-911587-53-8

All Rights Reserved. Copyright © 2021 Philip Burton. No part of this publication may be reproduced or transmitted in any form or by any means, without permission in writing from the author. The right of Philip Burton to be identified as the author of this work has been asserted by him in accordance with the Copyright, Designs and Patents Act 1988

The cover design is Copyright © 2021 Camilla Reeve
The front cover image is Copyright © 2021 Lynn Bushell, lynnbushellart@gmail.com
The photo of Philip Burton is Copyright © 2019 Laura Burton

A CIP catalogue record for this title is available from the British Library.

Dedication

I dedicate this book to the hundreds of children from whom I initially learned how to write poems, to my fellow-poets at Ribble Valley Stanza, fellow-writers at Clitheroe Writers Group, and to the Burnley and District Writers' Circle for their support over many years. Thanks too goes to Copland Smith for eight hundred hours of WEA creative writing courses and for the constant poetic encouragement, and to my family and friends.

Acknowledgements

A FINE LINE was awarded 2nd prize in the Ilkley Literature Festival poetry competition 2013. The judge was Gillian Clarke.
ANOTHER COUNTRY won First prize in the 2018 Sandwich (Kent) *Poet of the Year* poetry competition.
ANTRIN LICHTNIN is published in Stand Magazine #September 03, and is included on a CD titled *Falling*, published by Ann Wilson, in Barrow, 2006.
BARBERRY is on social media via *Back from the Brink* to support conservation work.
BLUE GRASS BLUES was published in the Commonword / Crocus Books anthology, *Peace Poems, in November 2002.*
CANAL COUNTRY was awarded a Distinction, and also the Senior judge's discretionary prize, in the Mere 2017 poetry competition organised by Pennybank Writers of Mere. The Senior Judge was Dawn Gorman.
DUNSOP VALLEY SPELL was published in The David Jones Journal #2008.
FLOOD won First prize in the Sentinel Literary Quarterly poetry competition, August 2014.
GUEST WORKER was published in Brando's Hat *Tarantula Publications*, Issue 11.
INSCRIBED ON A SUNDIAL was published in Dream Catcher #27, 2012.
LATE AUTUMN WHISPERS, in a previous version, was published by The Piedmont Literary Review, Santa Cruz, California, *Vol. 22, No. 4,* and it appears in Manifold #42, September 2002.
MEN IN BLACK was published in the Commonword / Crocus Books anthology, *Peace Poems,* in November 2002.
RAVEN IN THE MARGIN won the BARN OWL TRUST poetry competition, 2017. The judge was Mary Woodward.
SKYLARK was published in *poetry monthly press and graphics* #119, February 06.

SLOE TRAIN was published in *Iota poetry quarterly* #53, 2001/1.

THE ECO-NEUTRAL REVOLUTION was published in The Affectionate Punch #10, and in Pennine Ink #23, February 2002.

THE EVICTION was published in Stand Magazine, # September 2003.

THE HERRING GULL – an earlier version of this poem appeared in Moonstone #90, May 2003.

THE LAST RESORT is published in Rain Dog, #January 2005.

THE NEW FOREST CHILD is published in *Work Town Words* edition 15, *Wild* theme.

THE NEW OASIS is in Rain Dog #2, September 2000.

THE PRIZE received a Commendation from *The Barn Owl Trust Wildlife / Conservation poetry competition 2011-2012 and appears in the anthology.*

THE ROOST received a commendation from The Kent and Sussex Poetry Competition 2009, judged by Penelope Shuttle.

THE SAND GARDEN was included in The Swansea Review, #21, August 2001.

THE TREES ARE DOWN was published in Breathe Poetry Magazine, #June 2002, and was also published in the anthology, *A Trail of Words* (inspired by the Forest of Burnley), Mid Pennine Arts, April 2002.

VIEW FINDER received a Highly Commended award in The Welsh poetry competition, 2015.

VILLAGE CELESTE is published in Manifold Literary Magazine, #42.

WHEAT SHEDS A TEAR was published by Work Town Words #10, TEAR-themed edition.

WHIRLWIND WOLF is in Skald #10.

Contents

Gaia Warnings	1
A. Dismiss your missiles, children	3
Europium	4
Range-Finding	6
Men in Black	8
The Unknown Soldier	9
A Sleeping Promise	10
B. Go back and take the path less travelled	11
Pendle Hay Ride	12
Family Getaway	13
Slow Train	14
The Sand Garden	15
Ode to Sheep	16
Blue Grass Blues	17
Dunsop Valley Spell	18
Canal Country	19
Inscribed on a Sundial	20
The Roost	21
Guest Worker	22
Long postcard from a high place	23
The Allotment Chair	24

C. Don't be surprised you feel estranged	25
A Fine Line	26
The New Oasis	28
Forbidding Pasture	29
The Prize	30
Swan Down	31
Another Country	32
Living with trees	33
The Trees Are Down	34
The New Forest Child	36
Nor Wood	37
Barberry	38
The Beachcomber	39
The Last Resort	40
The Eco-Neutral Revolution	42
Village Celeste	43
Raven in the Margin	44
D. Listen to the loudening sob of creatures nestled in the fog	45
Wheat Sheds a Tear	46
Skylark	47
The Herring Gull	48
Flood	49
The Dry Spell	50
If, Come The Hurricane	51
Whirlwind Wolf	52

E. Tip your ploughshares into fiords	53
View Finder	54
The Eviction	55
Antrin Lichtnin	56
Rebel Ox	57
Late Autumn Whispers	58
Philip Burton - Biography	61

The forests are disappearing, the rivers are running dry, the wild life is exterminated, the climate is spoiled, and the earth becomes poorer and uglier every day...
                    Anton Chekhov, Uncle Vanya 1899

Gaia Warnings

Don't be surprised you feel estranged.
Listen to the loudening sob
of creatures nestled in the fog.
Learn this lesson. Mark my words.
Dismiss your missiles, children.
Tip your ploughshares into fiords.
Go back and take the path less travelled.

Reconnect with poems daily
by Charlotte Turner Smith,
John Clare, Coleridge,
and Joanna Baillie.
Is the Natural world today as rich?

A.
Dismiss your missiles, children

Europium

1.

There is an element of surprise
that, in nineteen oh one, they chose

europium to name a new metal;
not hard for us to see the parallel

with Europe back then – hugely volatile –
needing to cool, congeal.

Had Europe simply Europeanised,
become a single element at that time;

had its fragility been self-observed
and the barrels of guns repurposed

as magnificent city organ pipes;
had it furnished scientific minds

with a vital compass of compassion
so, when chancing on unstable atoms,

instead of breeding cosmic mayhem
researchers would have re-interred them

for fear of unleashing evils to come –
but no, we're saddled with *plutonium*.

2.

Old radioactive atoms turn leaden –
as do spent powers like Britain.

A world of exclusively inert nations
might constitute a more peaceful place

but *plumbum* does a rum thing, alas,
oozing down the stained glass

a full inch every thousand years.
Indeed, all history one day disappears.

The UK, peering past its hospital mask
yearns for a long-unfinished task

to resurrect. And this last firework can,
appearing to be dead, burn our hand.

Range-Finding

after Robert Frost

1.

They unlimbered cold cannons -
their talk festooned with procreation,
God, shit and swallow-tail rags

of ordinance jokes – eggshell thin.
All day long, the quartermasters
ticked buff boxes and men fell in

to learn the range, the elevation
that would wipe the tree-line flat
as the playing fields of Eton

the Captain said. Then a pipkin
of stew was gone round, and rum
for the lads to kill the din.

The sparrows were the first to fall.
Shock waves took them down the sky
as if they'd not been there at all.

2.

I am a white owl. Witness this.
I revisited my young throughout -
went about my fine business

till expelled from the hooded land
by explosive coughs of gore and sap.
Now I roost on a rookie's hand

who fired blood in short bursts
his limbs flapping along the margin
as though to fill a few fat bells

of uptrod flowers before the end.
There's not blood enough within me
for such noble gifts - minuends

of less and lesser drops of red.
Why do men do such harm and good
far from barn and potting shed?

Men in Black

He's felt the apologetic lurch
with which a plough
unearths a high explosive shell

but in a corner of a foreign field
that is forever dangerous
Luk shrugs, finishes the furrow.

He damps down the engine,
walks the long stir
of turned ground and bullish gulls.

He pauses, studies, frowns, turns
toward his rig,
at peace among the milling birds.

He pours coffee, unfolds breakfast
on a gingham towel.
Crisp dark bread. Sweet tomato

and today's gift from the hive.
His wide hands
cradle a muddied mobile phone.

He wipes it quiet, shields his eyes.
A NATO plane
glides between Armentieres

and the careless-lidded honey
of landscape
slowly finding its ancient level.

The Unknown Soldier

Sightseers come to stake their claim
as though reviewing next door's fence,
their minds on Hammersmith's traffic flow,
or Saturday's big promotion game,
the air of terror – price of vigilance.

But the church is dumb, except its bell
and a mum saying, *Sorry for my brood*.
Peace – being too long stabled with the dead –
something dramatic ought to happen here
at this fallen envelope of stone
perimetered with weary poppy feet –

The Angel of Mons perhaps
or the man himself, promoted pawn
of the sacrifice, dying to be the decent-
minded youth who'd murder a beer
and, shot of this musty chequered zone,
enjoy an hour of footie in the street.

And so we shuffle blankly out
spent as cartridges. Big Ben bound
we steel ourselves for one more shout
at seeing London in the round -
its peace campaign on the roundabout.

A Sleeping Promise

*a performance of Sleeping Beauty by Tantsy Na L'du, a
touring Russian ice-dance company, in Ramsgate in 1959*

The iron curtain disarmingly
lifted its skirts and drew us warmly in.
Tchaikovsky's waltz caressed us.
The ice, ungrooved as yet, glowed
with a giddy scoop of lights and halos.
New delights of spirals, sit-spins
and breathless upright Salchows
banished all thoughts of Cold War.

We sat convinced that Carabosse
would not harm the teenage star,
would – with a change-foot spin – kiss
and whisper *Pomirit'sja*, making peace.
The needle's poison would dry up.
We'd all embrace, on the rink, barefoot.

B.
Go back and take the path less travelled

Pendle Hay Ride

I'd like to ride on an old farm cart.
I'd lie in the sun and wait for the start.
I'd fall asleep on the spice clean hay.
I'd wake as the driver pulled the rein

to say, *Gee up Frankincense!*
from Higham to the village of Fence,
up the back lane, along the top road
to Old Read, Trapp Forge where we always slowed;

then White Hill, Well Wood, Black Hill,
Chew Barn, Nutter Barn, up past the school,
Pendle Nick, sand pit, and Coffin Stone,
Wymondhouse, Audley, and Cold Coats,

Tarry Barn, Ivy Cott, Standen Hall,
Clitheroe town, our very last call.

Family Getaway

I fidgeted all the ditch-long journey
envious of my sister's headlong sleep.
After the wringing-out of sandwiches –
and top-heavy march in heavy top fields

with canvas that had fought at Mons –
we settled on the forest edge
stiff and cold as the pines themselves,
in waterproofs heavy as Cornwall.

I had packed the usual suspects –
dismal diary, bare-spined Dickens,
liquorice lace – to tie the journey's ends.
Sealed against toads, we unfurled

our brief beds, wet in our sacks,
our eyes rainstuck, touching home
by mercy of dreams. Then the going down
for milk, carrying dawn on our backs

through cow pat fields, like sappers.
Then a powwow round the primus
making puddle plans and stroking knees
like poultices, scouting for songs.

Narrow, awake, among the dark
I sensed the dust heart of the conifers.

Slow Train

slow sloe train
well-named, pauses often,
often nowhere near a station
often to let roads through
or cattle, or fox to safe haven

or give a lift to a rambler
who looks a touch tired
just for conversation
with one of the five guards
or vacationing signalmen
who enjoy the train's pace

or visit the driver's daughter
who ties her blouse on sticks
to flag us down

or cool our feet in the river's
murmuring tones
or write poems about sloe
liqueur (made with gin
in a still on the slow sloe train)

The Sand Garden

I unearthed an edge to the new ground.
Mineral with animal connections:
a gravestone larger than death,
a standing stone grown tired, sunk
back to loamy bed, latent, duvet-deep
for me to find.
                But rather than disturb
I leave a space in the Escallonia,
welcome the sand as it idles through,
put shallow things here. Radishes.
Driftwood brushed by starlit oceans.
Shells combed pale: Peppery Furrow,
Thin Tellin, Cockle, White Tortoiseshell

insubstantial as the moon

Ode to Sheep

I like being ignored by the watchful eye
of a flock of sheep on their daily grind.
What silent airy landscape, almost sky
they nearly inhabit. Where do they find
the stimuli to know they are not dead?

Sun-rich buttercups and pot-hole smells
can only feed a five-minute ecstasy
and yet they look alert and interested.

Perhaps their busy minds store details
for a book 'Grass Management *Looks* Easy.'

Blue Grass Blues

I'll live in grass that's thistle-blue,
heave buckets to flush the john
watch the wide Missouri
as it ambles broadly on.

I hunger for hootenanny
where night is hot and long
and belt-and-braces canny
men to tell me when I'm wrong.

I hanker for candy-crunch washboards,
and banjos like peppery spoons.
I yearn to roll on cobs of corn
with cousins in tumble-down barns.

I yearn for young Virginia
with a rip in her cargo jeans
a hickory lap-top processor
and her rocking-chair that leans.

I'll live in grass that's thistle-blue,
heave buckets to flush the john
watch the wide Missouri
as it ambles broadly on.

Dunsop Valley Spell

The purple flags of Foxglove
and Creeping Thistles rise to shine
above browbeaten bracken.
A skeleton track tells, in runes,
of high trees brought low.
Soon hillsides will be broadleaved –
smart cash has fled from Pine.

This is shooting-party land.
The needs of Grouse, at any cost
are met. And yet the river
chooses its own bed, and how
the day's Oystercatcher-quarrel
and wind that wings the willow
pricks the casual visitor to life.

A vaccary wall's tight stack
of white-watered river stones
holds the past in a sentence line.
Monarchs still come here
making a progress, breathing;
the last to shoot the wildlife
was Richard Coeur de Lion.

Canal Country

I'm the narrowboat chugging my weight
from berth to berth. I'm wild-oats.
I'm the rounded robins who sing all year.

I'm two arched necks of the bridge.
I'm the swan's prolonged liquid glide
along the canal, smooth as daylight.

I'm the pound, the watertight caisson,
the glad lung inhaling its timely spill
to lift the boat and with it the world.

I'm the windlass key, pinion and rodding,
the paddle (with pawl to stop me dropping.)
I'm the weathering of the winding gear.

I'm the balance beam that gives the gate
an easy arc. I'm the stubbing post that smoulders
in a rope's embrace. I'm a scent of canal —

wild ginger, wakerobin, marsh speedwell.
I'm the eight white petals of a bloodroot
near Milepost 184. I'm Gill-over-the-ground.

I'm the Navvy, twelve cubic yards done,
pricked by exhaustion, resting my iron spade
for rapturous robins to hop on.

Inscribed on a Sundial

*at the National Wildflower Centre, Knotty Ash –
a found poem.*

twelve mid day Goat's Beard closes
one p.m. Scarlet Pimpernel closes
two p.m. Hawksbit closes
three p.m. Bindweed closes
four p.m. White Waterlily closes
five p.m. Common Poppy sheds its petals

six p.m. Evening Primrose opens
seven a.m. Dandelion opens
eight a.m. Mouse-ear Hawkweed opens
nine a.m. Field Marigold opens
ten a.m. Common Nipplewort closes
eleven a.m. Star of Bethlehem opens

The Roost

Brookside Nurseries, Rossendale

Rustic, rust free, underblown –
in the way that woodsmen handle the task –
the roofs aren't the usual Toblerones
but reflect the heights and habits of plants.

The freehold is owned by Muscovy ducks
and there's Garden Centre periphery
where *the unbilled* visitors have to tut
their troubled way through volcanic ferns.

Ah, *Pennyroyal*'s nice, but it soon dies.
Would you say pansies are right for a him?
And where in pots are the *Persil fries*
that Nan applies to her rheumatism?

Between the Malay terracotta –
can you see, can you see, white and umber,
one eye a Flanders poppy, one not,
a bird on a break from the mud pool?

Did you know they're over from Brazil?
They have this yen, folk there, for soybean.
So she can't roost – habitat if you will.
She's taken off now to the old Ash tree.

You forget that she has the wing-power.
And they all flew last year. To Baxenden.
Licking their long claws, they returned
with a lad in a pantechnicon.

Guest Worker

The moon is amplified by sunflowers.
Moon-clouds waver above shadows.
The earth's underarm is sweet. Showers
all day have topped and tailed the meadows
framing the buttery stone of Cheltenham.
Along the oatey-mortar-and-honey walls,
along rebel areas of wild marjoram,
beyond the egg-white gates and lintels
strides the urban fox on his quiet quest.
He yawns in contempt at my mock whistle,
runs to paw the blood-smells, is my guess,
behind the Co-op. Then a black bristle
of garden sweeps him, inheres him.
I turn my collar inside, like one wet ear.

Long postcard from a high place

before Cliviger's
permed houses
and gentle valley logic

but after the axed pines'
lost reflections
on Water Village reservoir

drive higher –
steer left to the lay-by
of diverse absences.

Largely untrodden –
unsung as a packhorse
ghosting through night's tunnel –

I am Crown Point
untouched by contours –
a notional place

equidistant
from Burnley, Bacup
and the sun as it sets

moor-grassed
on hard shaved
cool partings of hills.

The Allotment Chair

Jed's plot was a throwback to Eden.
Dog strollers broke stride, workmen softened.
Sunday folk watched him charm turnips
from early winter beds, shake his hips

at girls. And he'd give a deft display
to show that digging's about wrist, and say
that a thousand pounds drive a spade's haft
but a soft forearm and a gentle shove

will turn the clods, even in quagmire.
There was lightness in him; he'd not cease
till catalogues refrained from new varieties.
But the clanging of church spades

told of an elder's death – and it's Jed's turn
to lock horns with the allotment chair,
lend his plot to younger blood, and laze
beneath his laurels, turn seasons into days.

He toys with orchids, cress, Bonsai,
the raking of castrated leaves toward the fire.
But most he sits so long he's taking root
awaiting time and tide like King Canute.

C.
Don't be surprised you feel estranged

A Fine Line

Kerrigan whittles the timber –
allows a leeway – a safe margin –
leaves the cabinet man to plane
"trim as Mary Pickford" – goes over

to wrestle and steer home
the spindles to the cresting rail,
coax the stretcher, dove its tail
and sail his hand up the form

patting the five point low relief star,
praising the craftsman at his bench
who chisels a personal touch
on ships furniture beaux arts.

On the Antrim hills, fog is sand
and the horses choke and chafe
but the cart is hitched, made safe.
Kerrigan has a currier's hand.

In rhythmical joy, the men hand down
seat on seat, neat as Morse Code.
The wagon creaks like a whale's bone –
a din that his cheery song drowns.

Donning his vest, he stops to invoke
Saint Coleman, give the chairs a scrub.
He heaves them onto the ship at Cobh
sleek in their coats of shellac.

He sets the deck chairs stern abreast.
Then, being kindly, twists them clear
of the sad eye-line of Heartbreak Pier.
He ponders, turns them back once more.

The New Oasis

The tarmac appears permanent today.
But 'road-metal' is just weakened stone.
Though early sun shines it like a railway
the sub-soil's only marking time below.
On the estate corner churned by streams
of diesels clumping the odd wheel over
I notice for the first time: a fault, a dream
of what was once there: a galaxy of clover
lit by a buttercup supernova.
I walk on hurriedly. The oasis
can't persist for long, won't recover;
one random swerve of an articulated chassis…
Where I turn for home the road swoons
in a cul-de-sac, gives birth to a garden.

Forbidding Pasture

The field is hemmed by hawthorn,
wet stone, and the thin gesture of wire.

Each fresh season I'm drawn
to dig my old bones out of the divan
and ramble the Lottery-funded footpath.

I peer through the padded curtain.
Hot-day gusts pollinate my eyes.
And nettles? All yours, butterfly-man!
I stay and pay the rent or scratch the nose.
September. Cow-pats reek the grass.
Bramble leaders knit a noose.
Or the timbre's wrong. Soon frost
will lock me home, a Jonah with a belly,
and only Mrs Fox will test
the right to roam.

The Prize

Between the night and the village
sodium glow creeps unwarranted
along the breastworks of a ditch.
A chaffinch nest, pierced, dislodged,
lies on the desolate brim of the road
betrayed by the gross modern rage
for chain-saw management of hedgerows
to ease the Grand Guignol rampage
of the car. The villagers arrive for Easter,

swerve to avoid the tumulus nest.
A polecat moon paws-back the cloud's edge,
looks in. One fledgling left, on tiptoes,
cries in silence. No replies.
The cuckoo village quietens as it squats.
Day by day, carefully sewn moss, roots, grass
unwind, buffeted by cars which shuttle past.
A figure sweeps the parking lots.
Today's The Best Kept Village prize.

Swan Down

Fuldatal, Germany, December 2020

Overhead lines brought down a swan –
dawnlit pages flicked loose from a bleached folder
posted through a letterbox. Strewn.

There had been two. The one spared,
the bereaved, the outliving soul, undertook
a signature plunge, an ungraced blur

of fiercely knotted wedding dress
down the aisle of sleepers. Though sad as chiselled salt
its majesty covered the white ballast

and held hostage the ribs of the railway track.
The trains caught the mood, cancelled themselves
from miles away, sensed the ache.

Grief was allowed the commuter too.
Something was gone, more than mere timetable.
The station clock was heard to tick.

Another Country

if one keeps on walking – **Soren Kierkegaard**

Hike from a French village. Its name
is crossed out on a sign where *maisons* end;
no ribbon development, *merci*, no dullsville

fringe; *la bourgade* is out of the frame
in a few country strides. English hamlets
drill outwards into the wild, an overspill

that cakes the ancient fund of country lanes
with tight little builds. Undone Green Belt
melts, trickles, cannot help but seal

the walking routes. We saw with chains,
reconvene the wood as high green fences.
Parking ramps cut short the ditches. We kill

the thing admired. People will embrace
clipped landscape, hungry for what was held
to be a good picnic spot now concealed

by a Flemish wall. The road hiccups over a bridge
and, there, the many loam-deep scents
of countryside. And it's alright. A bell peals.

Living with trees

*A poem summarising the 2020 research of Aki Sinkkonen,
University of Helsinki*

It is with delight we found out today
and the proof cannot be stronger –
that a little forest provided by school
out where the children daily play
ensures that those children are, as a rule,
healthier and live longer.

The Trees Are Down

on helping to destroy an olive grove
to make way for new cotton fields

At harvest time the white rising air
of cotton fields draws the musk, of grapes
and ciderous apples, through the grove.
Today, deep in fallen olives, the labourers
question why the trees get no reprieve.

Sharpened by gloom, a farm boy climbs.
The branches weave their gentlest spell
like a dog with a violent master.
The bough stoops, begs for sanctuary:
each leaf with its own silver certainty.

The timber gives way.
We're showered with boundless drupes
and the venerable dust of old pollen.
The Dryads weep for this 'loss of balance'
which we carry everywhere; first we shout

then, in the same choking breath, we laugh
and leap like frogs among the fallen
limbs of wood, and pluck, with iron fingers,
warm olives to keep them from the fire.
Only the tree roots are staunch, till sprung

from the earth, uprooted by hydraulic digger.
A thousand years of quiet fertility
reduce to smoke in the gloaming air.
The breeze blows harder in the absence of the grove.
One by one the farm shutters close.

The New Forest Child

Where is the forest where children play?
the new child gasped. *Is it out back?*
Do you grow moss, blueberry, bay?
Are there buckets and spades for needles,
cones, acorns   chips of bark?
In the summer weeks of Stag beetles
do kids stay   and watch till dark?
Why have you put your forest away?
Please get it out. It's for children's play.

Oh no, we are terribly   modern.
Hardstanding and gravel yards are bare
to ensure that each child stays within vision,
is clean, and learns not to care.

Nor Wood

This green plot shall be our stage

— Shakespeare, A Midsummer Night's Dream

So much happens in the forest. Not yet grasped our role
in the overarching script, we rip the playbill, fire the scenarist.
Quercus robur, English Oak, palace of variety, hosts a company
with which we should audition (our playhouses now dark):
badger, squirrel and deer in the royal hunt of the acorn.

In the gods, pied flycatcher and marsh tit call and re-call.
On the skyline, a tree walks a tightrope quite unobserved.
Balancing-pole roots steady the trunk, deflect the storm.
Roots underpin the fields along the woodland edge, *sub rosa*,
stitching delicate loam in place, drinking rain down.

Forests stand on a world stage: elders, thinkers, planet healers.
Each twig waves a maestro's baton. Bow low to the boughs.
Leaves perform as vibrant choirs. A landscapist may paint
a sylvan canopy in one green stroke, but each blade's a one-off
just as we are – a compound of spring and of autumn.

Bring down the curtain on timberbeast and his wolfish
plaids, racing axe and buck saw. Time to hug a tree,
reach out to heartwood, not for timber but wisdom
not for fruit but fruition, not for latex that gentle milk
and not for willow bark tea, but to understand authorship.

Barberry

It certainly was there but
the Barberry carpet moth hid from our eyes,
immersed in curved triangulate wings
feather-clouded in swirls of grey
traversed by a brush-flick of twig
and two dabbed depictions of leaves.

Unfazed by us, it would be there
but we took its sturdy food plant away.
Its rust fungus, we figured, blew on
to the wheat. Berberis burned all day.
Purple barberries roasted and steamed.
Towpaths danced ghost grey
with cuticle dust and bereft wings.

When a new strain of resistant corn
came on line, the long guardian hedges
were gone. Pragmatic gales wore
the rootless loam to dust in the drought.
Floods dissolved canal-bank edges.
Carpet moth, teach us a gentler walk.

The Beachcomber

for Miss Tandy Richards

She had a way of living her life
in the middle-distance.

Pockets as large as her blanket coat
would harbour a cubit or two of
        sea-blue anchor rope,
evocative white driftwood of a less acidic sea,
a deposit bottle bearing a note,
time honoured additions to the sculpture of shells
that she found as divine as
        Ulva weed with its truffle smell.

Now her granddaughter chugs the strand
in a machine – a sandboni – harvesting marine debris,
filling council wagons to overflowing. And land-fill isn't free.

Along the Sound, the shallow bights,
the towering duffel flaps of hungry coves
buttonhole shoals of baby liners
polymer pots and nano-detritus
and guide them to rest in a swaying heap
beyond the arms of the tide.

The Last Resort

Glifada, Corfu

They say the caterpillar tractors
came from landing craft. No-one saw.
And all the shapely thigh of hill
was shaved of Aleppo and Black pine
in a single night. The drivers fell
drugged by the resin, they say.

One year ago you would have looked
at olive, oak, the cypress, the carob
of the brae, rearranged softly
year-on-year in natural regeneration.
And the whispering white poplars
told of no settlement by man.
You would see the orange speckled clan
of salamanders taking early sun,
the plane tree near the stream,
the tree-of-heaven palm
in curved cascade between
        the darkness and the day.

Ulysses, shipwrecked on this sand
slept in his own arms. Here he saw,
in a porpoise's eye, the Atlantic sunrise
filtered through Herculean slabs
of Africa and Spain, liquefying gold
from a stadium of skies.

Sternbergia still reflects the sun's niche –
its low yellow flowers are a sealed knot.
But I miss the white petals of the sea squill
on a high September stalk, their multi-level
perfume, and their gentle way
with hours of storm, and hours of still.

The Eco-Neutral Revolution

There's not a single jot of wood
in a computer,
nor in a computer stool.

There's no deforestation
in manufacturing a 'Play Station.'
Compact discs are chic
and need no mahogany or teak
to hide in.
Keyboards don't use 'boards' at all.
You don't need to mow down pines
to make a modem.
When you download
a document
no tree is touched by the event.

I'll go further,
maybe into Thurber country:
in all the Internet's electronic
gossamer
there's not one interaction
with a tree,
not one connection
with chain-saw-economics
or slash-and-burn, or land reform
or forestry, or jungle husbandry.

In all its megabytes of memory
only pictures of a tree,
just electrons dancing
where the leaves used to be.

Village Celeste

Roads quarter the common. Evening drags
at dimpled stone. Residual clock-hands
on the tower proclaim that it is midnight
or, if day, noon.

All is closed save the graveyard.
The stock of village life (the pub, the school,
the general store), hollow megaliths
among the solid failure of the corn.

Above a lodge a weathervane is bent
immobile, as if one sign of change
would summon Death, disturb the warden.
A window droops in ivy's dull mask.

The wine-cellar door is steeled with nails.
The famous kitchen garden, umber-lit
encroached by pines, shows only remnant
mutant cabbages, and dry cider-traps.

No meadows of tawny, tan and grey;
only standard gold, and modified rape
to come, and the grubbing-machine
in the long hedge of a child's laugh.

Footnote: The last lines are a reference to the poem 'The Harvest Field', by W.R. Rogers, in which appears, 'the scythe in the long grass of your laughter.'

Raven in the Margin

Hraefnes Geat ... Raven's Gap ... Ramsgate

I look for traces, floated feathers,
clay-pipe bones, outline of a razor clam.
Each day the tide sets out its wares.
One raven has left a bone – an ilium –

long hidden, now exhumed, eased
from time-sifted sands to a cartwheel life,
shifted by storms, thinned and cleaned
of unkindness. Here's Sybil, a fishwife,

with a raven so tame it's absurd.
Syb's last of a line of women – piercers,
toppers and gutters of herring – who fed,
age after age, raven and seabird.

Industry triggered a fish glue works'
sulphurous whiffs, plumes of ammonia,
quicklime in white heaps, coal from Dover,
sand to make glass from washing soda.

No More, says the feathered beast,
will the tides rinse as clean as you suppose.
The fish are dead, my family up and gone.
The sea feeds on your conspiracies.

D.
Listen to the loudening sob
of creatures nestled in the fog

Wheat Sheds a Tear

I'm bread-scented
and rise naked from one seed
in the subsoil rut of an over-managed field.
Cryptaesthesia is my advance ticket out of here–
a psychic connection to the warming atmosphere
smoking above, to the heaving life of the unfarmed fringe,
to the ripped victims of the plough, those crushed beneath
tyres.

I sense, below, the leaden past of petrol tractors, the acid attacks
of envenomed rain, the ham-fisted doling of chemical fixes,
the unrecorded death of undiscovered single cells –
billions in each gram of soil – the knowing
next winter will widen the river
and rip-tides dissolve
all scent of life.

Skylark

chanter alouette

swift as old age
skylark died

skylark died –
we are dumb

skylark – Beethoven
high as hot helium

no better call
on people's time –
to hear, above all,
and all day long
skylark song

when skylark lived –
we were deaf
skylark died –
we are dumb

The Herring Gull

When days were young I'd rig a hide
out on the rocks, at the heels of the ebb tide
and be with herring gulls. Summer years
were given to the task with no return
but the ocean and to find my calmer side.

Unwinding from the bobbin sky the birds threw
high tuba calls without sacrifice of altitude.
Forty shades of green meat failed to feed
a single beak. Gulls dipped and rose as though
a glass floor kept them.

One preternatural day in the mid-oven of noon
tamely on stiff skin legs a herring gull came.
A vestige of line strayed from the loud beak
and, no question mark, a fishing-hook
had snared her craw.

Something must have said, *That man there –
he's mad, but means okay*. I didn't ask the vet
what procedure had to say but watched the claws
as the steel was drawn away.

When I'm down or ill, I stretch my arms and hear
a trumpeting of skies, and strong wings rising.

Flood

On the winter-caked lawn, snowdrops arrived first
like a spill of blanched almonds (or dew-drenched confetti).
The flood overtook them – not speedily – but worse –
despite the causal elements of wind-rich tides

and denser-than-usual rain – it came as a dead thing –
no pulse – cold and squat as acres of window glass –
a super-cooled molten mirror rising and swallowing
garden steps in small steps, sobbing through air-bricks.

Imagine the most unwanted kiss: a fierce tongue
which penetrates the loose fabric of your being.
Not only will you feel unclean. Pleuston –
a floating mass of microorganisms – enters in.

The rustic solid oak sideboard which made you
proud, grounded and secure, can't navigate
the narrow stairs, stands foursquare in the ordure.
The world you thought was your dominion

is a wetland swamp we merely borrow.
After a stay of three weeks, even if the floods decline
and you search, less in expectation than in sorrow,
there will be no snowdrops, no lawn, no garden.

The Dry Spell

Wirksworth Cricket Club 1963

Such white that year! Limestone breeze
dimmed the cat's-eyes road, the trees,
speckled the wrens' eggs, dusted the field,
rotted each stitch in the canvas marquee.

We'd people to go, placements to be,
those dread digs and University.
We stowed the kit and kissed the crease
and promised that we'd often meet.

Whiter than the ancient screen
the pitch held footprints hornblende green
dry as any grass has been,
drier than dreams, than fresh green

departing men, or angels suddenly,
lifted by the moment, shaken free
of close friends, already empty,
eyes made white by lightning sheets.

White water canvassed down between
the lost pavilion's creaking seams
bowling-out the pastel scene
knocking the bails off our late teens.

In winter snow I often dream,
transported to those white streams
that downed the tent in that white field,
when I was young and at my ease.

If, Come The Hurricane

If, come the hurricane
you can sit in your tent
and smile while a caravan
somersaults past your door-flap
you have a touching faith
in the breaking strain
of nylon. Either that
or you're a mountaineer.

Whirlwind Wolf

It has run ten seasons to find fresh meat:
the wolf-wind that's at the back door howling.
It'll cook our goose on Beaufort Mark Twelve,
though the drainpipe is painted with antifouling.

The weathercock's gone with a splintery snap.
Molars are rending the heater cowling.
Tempered steel? Soft margarine, for
the wolf-wind that's at the back door howling.

The fat of the land greases her lung
to inhale the plumbing, and vomit the towelling.
The guttering's gone to her guttering gut,
though the drainpipe is painted with antifouling.

The happy home's in her maundering jaw,
and the flesh of the brick is dissolving.
It's all she-wolf round the front patio:
the wolf-wind that's at the back door howling!

She's quaffing us with a brandy-snap wine.
She's dissolving us up in her bilious fountain;
we're drawn up a straw, like water and gore,
through the drainpipe we painted with antifouling.

Now she's flying back to Gran Canaria.
Our lupins lie crushed. Our final pooch is growling.
The thistles lament. The bare foundations whistle
the wolf-wind that was at the back door howling.

E.
Tip your ploughshares into fiords

View Finder

He referees the landscape. His whistle is a camera;
one blast and a militant shindig of spiralling leaves
comes to heel. The spluttering tractor, swaying farmer,
the random grazing shudders of sheep – all freeze
as one; the rural ebb and flow are quietly on hold.
We wait for the wagged finger of natural justice.
But the official cops out, walks away, drives home,
uploads tranquil images from his lauded digital device
which cannot detect the stalling engine, skidding boot,
the felled driver caught below the encumbered wheel,
the ewes reeling away, the twisted anthem of a shout.
The human eye serves a slack brain which as its goal
longs only for a hug, easy passage, home advantage,
peace in the dug-out, our perfect lads centre stage.

The Eviction

The long brocades of slumbering kale
meet at the cattle shieling ground
Dry washing bunts, tourniqueting
The tithe cottage is up for sale –
the talk in 'Little Queen' is all around
her going. But the clothes are hung again
The balefire yellow dress, dancing wassail
flags her staying to all the village round.

Irene clips, crazy-eyed, up the paving

The bailiff's men disturb a pheasant male
Their stiff boots talk of shootings down.
John Smith looks, then counts his pigs in

The glowering sky, the vanishings
meet at the distant cattle shieling ground
No house here. Just embers, quitting
the long brocades of slumbering kale.

Antrin Lichtnin

A tame idea to have, that wildlife
would hunker down
and – like model folk in the military town
behind the bridal white
neck of canal –
defer to the veil of spotless rain
soon to hide Ben Bahn,
Ardgour, far away Mull.

But no. Above the scumble of heather –
needling in
on covert wings
like Venus viewed through conifer
a prey bird plummeted
onto cold fallow like a blown kiss –
the great gold nape, apotheosis
of pride.

Then the ruck of talon on fur
pre-chilled in the stubborn glen,
the slow metronomic flap again
of *force majeure*

then to high Gulvain
through soft confetti Highland rain.

Footnote: Antrin Lichtnin, *Scots dialect,* 'rare lightning'. Which is what poets live for. (Hugh MacDiarmid's poem, 'A Drunk Man Looks at the Thistle').

Rebel Ox

Research scientists at Oxford University, Harper Adams University, and at Soulton Hall, found that traditional ploughing causes harm to the life of the soil.

Athena – goddess of wisdom – sprang
in body armour, from the head of Zeus
to teach mankind to penetrate the earth.
Ever effective, too, are parallax marks
in the field: right eye, far blackthorn,
middle distant tufts of bent couch grass.
Bush, clump, eyeball, make a straight line.
Surely the plough will always furrow true.
No protruding rock bothers the bullock –
though one bronze-dry sod of grass can lurch him
sideways for a moment, his bulk stilled.

But always the leather traces creak,
the mouldboard hefts, the ploughshare
bites and off he sets again to the lone tree,
the rippling turn, the elongated return slog
through subsoil, peat, elastic turf and rain.
Yet now he's bolted off, burst a blain
across the paddock, ripped the access road,
unseated rows of spring cabbages
and let his heaving flank collapse the piggery.
Perhaps he'll launch back across wild ages
to where land survives unscratched.

Late Autumn Whispers

The plough has turned, turned, turned
the life within the soil over to destruction.
The delicate webs are snapped by the sun
tangled by winds, even before we ken
they exist, name them, or know their purpose.

The fox flares her hungry nose. A speck of red
in the far field. A tired maple. Otiose Autumn.
The bilberry mist has rained its seed. The woods blaze
from shadow, though the funeral pyre's gone.
The fox-trails colden, stifled in lingering smoke.

The snow is slow to come. A shallow breeze
partly from the north, musses its paws
among dead leaves. Everything just exactly so:
meagre corn-meal in the byre: the harvester
greased and under wraps: the dead, quiet.

The lonely cartwheel spins the combing flakes
to brush its lean spokes. Leather straps whip
the alder-pegs on the flapping stable door.
The furrows narrow to a dotted line.
The tired vixen screams, and goes to earth.

The farm lies open as a night before a war.
"I come to you my love, to cover you,"
a whisper made of silence in the coppice,
the only-ever-promise made by new snow.
The grudge-deep ditch fills, is forgotten.

Next year this farmer will scrap his tractor
give his heart and hand to the no-till approach.
Fewer weeds will blight the harvest
and the wheat will ripen quicker, stronger.

Philip Burton - Biography

Philip Burton's love of readings and performance developed through life as an English and Drama teacher, Lancashire head teacher, folksinger, amateur actor, and as a poetry practitioner who, as *Pip The Poet*, has provided hundreds of poetry days for schools and for adult learners. He was born in Dunfermline and raised in Ramsgate, Kent.

Dyslexia was a problem for him as a child; he went on to study science at Edinburgh University. He received a commendation from Heidi Williamson in The Poetry Society Stanza poetry competition, 2020, for his poem on the theme of dyslexia.

Over three hundred and seventy of his poems have been published in literary magazines including *Stand*, *P.N. Review*, and *Smiths Knoll*. His poems have been widely anthologized appearing in anthologies such as *Best of Manchester Poets*, *Puppywolf*, and *Peace Poems Anthology* by Commonword / Crocus Books in March 2003.

In 2019, Philip held four First prizes concurrently in national or international poetry competitions: the National Arts Centre Jack Clemo poetry competition, 2019, the Horwich Writers *Hate Crime Awareness* poetry competition 2018, the Sandwich (Kent) Poet of the Year award, 2018, and the BARN OWL TRUST poetry competition, 2017.

Winning poetry prizes meant that Philip has met, and had the experience of reading on the same platform as Adrian Mitchell, Gillian Clarke, Penelope Shuttle, Michael Schmidt and Kei Miller. He arrived at The Green Room as "a new voice come to Manchester", and appeared in 2000 on the same bill as Adrian Mitchell at The Burnley Arts Festival.

Philip attended a University of the Third Age poetry appreciation group in Todmorden for three years and is a member of Clitheroe Writers Group and the Ribble Valley Stanza of The Poetry Society. For six years he has judged the Burnley Writers' Circle annual poetry competition, and was appointed as Honorary President of that organization in 1919. He has given a number of workshops at the NAWE yearly conference; and has also been a mentor with NAWE.

Previous poetry publications include *The Raven's Diary* (joe publish 1998), *Couples* (Clitheroe Books Press 2008), and *His Usual Theft,* (Indigo Dreams Press 2017).

Back from the Brink

The poems in *Gaia Warnings* reflect Philip Burton's passionate interest in wildlife conservation, especially that of Back from the Brink – one of the most ambitious conservation projects ever undertaken. Its aim – to save 20 species from extinction and benefit over 200 more through 19 projects that span England; from the tip of Cornwall to Northumberland.

It's the first time ever that so many conservation organisations have come together with one focus in mind – to bring back from the brink of extinction some of England's most threatened species of animal, plant and fungi. Explore the diverse projects on their website naturebftb.co.uk to find out more about the special species they'll be saving, the places they're working and how you can get involved and make a difference.

Palewell Press

Palewell Press is an independent publisher handling poetry, fiction and non-fiction with a focus on books that foster Justice, Equality and Sustainability. The Editor can be reached on enquiries@palewellpress.co.uk

Lightning Source UK Ltd.
Milton Keynes UK
UKHW011609160421
382096UK00009B/423